Advance praise for *i am here And You Are Gone*

In *i am here And You Are Gone*, Shome Dasgupta writes with a sparse and mathematical elegance, creating a system of symbols and spaces with which to describe young Jonas' growing affection for his friend Mary. Throughout the too-fast years of their youth, the two friends hurtle across what must feel to them like a lifetime, moving like two near-parallel lines approaching the same point, one set so impossibly infinite that they might never reach it together. Still their paths pull close, closer, so close that no matter how we might look we may be unable to divine what unsolvable span it is that at the last separates them--and so perhaps us--from each other.

Matt Bell, author of *How They Were Found*

From the start, I fell in love with this book. Shome Dasgupta takes on perhaps the most common of all subjects—love—and makes it somehow fresh and new. Stunning in its use of form, language and character, *i am here And You Are Gone* is a delight. The book and its author are remarkable. This is the first you may be hearing from Shome Dasgupta, but I can't imagine it will be the last."

Rob Roberge, author of *Working Backwards from the Worst Moment of My Life*

Shome Dasgupta writes about young love with a heartbreaking honesty and simplicity that transcends anything I've read in recent years. As we follow Jonas and Mary from kindergarten to senior year, we see a range of textual expression that is both innovative and wildly appealing. This book cracks powerful with youthful thunder, black echoes laced tightly with ephemeral innocence. Dasgupta's debut hits the mark, and leaves one.

Mel Bosworth, author of *Grease Stains, Kismet, and Maternal Wisdom*

i am here
And You Are Gone

Shome Dasgupta

Outsider Writers Press
2010

Published by Outsider Writers Press
outsiderwriters.org
in conjunction with Undie Press
www.undiepress.com

Book design and layout by Tim Hall, Caleb Ross, and Shome Dasgupta. Cover Illustration by Brent Watkins.

Please note that portions of this manuscript have appeared elsewhere, and thank you to the kind editors of these journals: "Chariot: Eulogy For A Paramour" appeared in *Mud Luscious* (Issue 7, April 2009); "Carving The Air" appeared in *Frame Lines* Magazine (Edition 8, June 2009); "Brief Encounter In The Museum" appeared in *Paperwall* (Issue 14, March 2009); "Parlez-Vous?" appeared in *Dogzplot* (January, 2009).

ISBN-13: 978-0-9763460-7-4

Printed in USA

Acknowledgments

Thank you: Mommy, Daddy, deep, Tabby, Andy Breaux, Stacey and Terry and Ada Grow, Luke and Lindsey and Sam Sonnier, Chad Cosby and Bianca Yoo, Clare and Aaron Dalbec, Brandon and Angelique Sonnier, Rien Fertel, Ryan Castle, Katie Frayard, Eddie Barry, Justin Bacqué, Casae Hobbs, Anu Gupta Desai, Jeff Distefano, Dallas Griffith, Matt Bell, Ryan Dilbert, Kristin Stoner, Patrick O'Neil, Andrea Pappas, Mark Maynard, Scarlet Tanagers, Molly Gaudry, J.A. Tyler, Tupac Shakur, Caleb J. Ross, Tim Hall, Pat King, OW Press, Tainted Coffee Press, Undie Press, Rob Roberge, Susan Taylor Chehak, Jennifer Ames, Dana Johnson, and Antioch University-Los Angeles MFA Creative Writing Program.

i
am
here
And
You
Are
Gone

KINDERGARTEN

Mary-Go-Round

Mrs. Jasker was Jonas' kindergarten teacher. She wore socks like the nanny from Muppet Babies, except they were purple and blue instead of green and white. She always looked like she had just finished crying and never talked much. She napped when the class napped, too. Mary and Jonas were on the playground, twirling around and around, until Hi-C came back out of their mouths. Mary's favorite cartoon was Looney Tunes. So was Jonas'. He asked her to marry him—she said yes, and he gave her a dandelion.

Mary Mary Mary Mary Mary Mary
Mary Mary Mary Mary Mary Mary
Mary Mary
Mary Mary Mary Mary
Mary Mary Mary Mary
Mary Mary Mary Mary
Mary Mary Mary Mary
Mary Mary Mary Mary
Mary Mary
Mary Mary Mary Mary Mary Mary
Mary Mary Mary Mary Mary Mary

FIRST GRADE

Mrs. S

Mrs. S had too long of a name—no one could pronounce it. The class just called her Mrs. S. Her nose was long too, like Pinocchio on a bad day. No one liked her. Her voice was high-pitched, and her neck was a mini-staircase.

She screamed at Jonas a lot when no one else was around.

"You are so dumb."
"You idiot."
"How are you alive?"
"I wish you didn't come to this school."

Jonas threw an apple at her on Teacher Appreciation Day. She threw it back at him, hitting his forehead. It swelled. He told Mom that he fell off the swings. Later that same week, during lunch break, Jonas took his teacher's purse and threw it underneath one of the teacher's cars in the parking lot. When she couldn't find it, she kept looking at him. He just smiled and continued to learn how to add numbers.

1 + 1 = Mary and me.
0 + 0 = Mrs. S needs to go away.
2 + 2 = My two favorite numbers to add.

The day after, she grabbed Jonas by the arm and told him to

"Go to Hell."

It was the first time Jonas thought about religion.

SECOND GRADE

High Heels

T was taller than Z; Z was taller than W; W was taller than P; L was taller than T, Z, and W; M was taller than W, but shorter than T; T, Z, W, P, L, and M were taller than J.

Jonas went to school one day wearing his mother's high heels. It didn't seem right to others, as they pointed and laughed and called him a girl—everyone, except for Mary. Mrs. Carrados asked him why he wasn't wearing his tennis shoes.

"Why aren't you wearing your tennis shoes, sweetie?"

Jonas liked Mrs. Carrados—she had grey hair and she wore dresses that always had some kind of fruit on them. Oranges were his favorite.

"I wanted to be tall like everyone else."
"You're perfect just the way you are."

She let him wear the heels for the day, but when he got home, Mom told him that he had to wear his shoes from then on.

"I want to be tall."
"You are as tall as you want to be—it doesn't have to show. Be tall from inside."

Jonas didn't know what she meant, but when he thought about it later, it was like some kind of spiritual assertion.

"I will be tall."
"I am tall."

Mrs. Carrados told her students that they were going to be her last class, and after that year, she was going away. Going away, Jonas found out later, meant that she was going to die from cancer.

"Capricorn?"
"No, cancer."

On the last day of school, Jonas told her that he wanted to be just like her. She said that he should be himself.

"Tall."

L
L
L
L T
L T
L T
L T Z
L T Z
L T Z M
L T Z W M
L T Z W M
L T Z W M
L T Z W P M
L T Z W P M
L T Z W P M
L T Z W P M
L T Z W P M
L T Z W P M
L T Z W P M
L T Z W P M
L T Z W P M
L T Z W P M
L T Z W P M
L T Z W P M
L T Z W P M
L T Z W P M Jonas
L T Z W P M Jonas
L T Z W P M Jonas
L T Z W P M Jonas
L T Z W P M Jonas
L T Z W P M Jonas
L T Z W P M Jonas
L T Z W P M Jonas
L T Z W P M Jonas

M R S. C A R R A D O S

Dirt Dirt Dirt

Thoughts:

If Mrs. S thought that Jonas should go to Hell, Mrs. Carrados should be the opposite, if there is an opposite. If there is a Hell, Mrs. Carrodos is the anti-Hell. But they buried her underneath—how do they make it to the opposite? Diffusion? Osmosis? Magic?

Maybe there are no opposites. Is = Is. Was = Was.

$$\text{Is} + \text{Was} + \text{Will} = \emptyset$$

Is it all a magic trick?

He doesn't like rabbits.
He doesn't like rabbits.
He just can't have it.
He just can't have it.
He doesn't like rabbits.

It's all a wonderland, after—he will be there soon to know the answer.

THIRD GRADE

In Between The Walls

Mary and Jonas were in different classes for the first time. She had Mrs. Beswith, and he had Mrs. Lorne. During each day, he would stare at the wall, wondering if she was doing the same thing. They would meet up during recess and lunch, but that was never enough.

One day, Mrs. Lorne was reading Peter Pan to the class, and as she was looking down at the book, Jonas got up and walked out the door and went to the classroom next door. Mrs. Beswith asked him if anything was wrong. He said he wanted to be with Mary. Mary raised her hands.

"Can he be with me?"

"I'm afraid not—he needs to be in Mrs. Lorne's class."

Jonas ran over to Mary and hugged her, before Mrs. Beswith told him to go back to his own class. When he went back, Mrs. Lorne asked him what happened:

"I'm lost."
"You're right here."

Mary and Jonas were able to spend time together when the class took a field trip to the zoo. It was just like the song—there were lions, and tigers, and bears, but Jonas was more interested in the otters swimming on their backs. It looked like they should have been wearing sunglasses and holding some kind of fruit drink that had one of those toothpick umbrellas. He turned to Mary:

"Would you like to go swimming?"
"Let's go swimming."

They climbed over the railing and jumped into the

water. The otters swam around them. The water was cold. They tried to swim on their backs like the otters but they couldn't move around much because of the surrounding rocks. Mary took his hand:

"We can't be otters."
"We can only be ourselves."

Not too much later, a man in khaki pants and a gray shirt shouted at them. He climbed over the railing and pulled them out. They went back over to the other side, wet and cold. The man said:

"You can't do that. That's illegal and dangerous. Where are your parents?"

Mary laughed:

"Mom is most probably working at the office right now. Dad? Don't know where he is? He took off a long time ago, before I really knew him. He didn't love us too much. Mom says he was a fuckin' motherfucker who loved his Advil way too much."

Jonas didn't know what a fuckin' motherfucker was:

"What's a fuckin' motherfucker?"

"It's a mystery to us all."

The khaki and grey man told them not to use such foul language, and Jonas thought about baseball, when someone gets a hit, but it didn't count, because it went on the other side of the line.

"Foul ball," umpires say.

Mrs. Lorne found them and apologized to the khaki-grey man. He went away, and Mrs. Lorne held their hands and took them to the bus. They dried themselves with napkins.

"So, to recap, we don't climb over the railings."
"Got it."
"I understand."

They joined the rest of the class and looked at the monkeys hanging from trees and jumping from branch to branch.

Mary was crying—she was thinking about the dad she never had. Jonas' stomach started to hurt. The emptiness inside made his head go blank and he bent over, before falling to the ground. He started crying. The class had moved on toward the giraffes. He saw the ghost of his second grade teacher, Mrs. Carrados. She floated around in circles, putting her hand on Jonas's stomach to help soothe his pain. It wasn't raining, but one raindrop fell on his head. Mrs. Carrados told him to go back to the bus and sit there until the field trip was over. Jonas did as she said, and as he walked back he tried his best to hold everything in, but even the slightest release gave him relief. His underwear was wet and sticky. He shifted back and forth in the bus, so that the stained spots weren't touching his skin. He still had more in him, but he just cried instead. This went on for about 2 hours, before the class came back to the bus. Mary came and sat next to Jonas, holding a slice of pizza.

"You don't want a slice do you?"
"I can't."
"Your stomach hurts."
"It felt like arrows."
"I can smell it."

Mary threw her slice of pizza out the window and rubbed his back. Jonas fell asleep on her shoulder.

When he got home, he threw away his underwear. Mom gave him Pepto, and he got a brand new pair of tube socks. After his system was all cleaned out, he ran around in the front yard, with his socks pulled up to his knees, pretending that he was Pippi Longstocking.

The Daily Chronicle

This past Tuesday morning, Julius Ewens was found dead in his apartment. The police say that his death was either a suicide or a result of an overdose, as he was surrounded by various prescription pills. He had a long record of drug use, theft, forgery, and battery. Julius was unemployed and divorced. He is survived by his daughter, Mary Mots.

FOURTH GRADE

Report Card

FINAL GRADES

Reading and Writing	A
Math	B
Physical Education	B
Science	B
History and Social Studies	A
Computer Skills	B
Conduct	B

Comments:

Jonas worked hard throughout the school year, paying particular attention to Reading, Writing, and History. For Math and Science, with a little more effort, he could have pulled out a couple of A's, but his homework assignments weren't quite substantial. More remarkable than his schoolwork, is his thoughtfulness to his classmates—always willing to share or help out. I can tell he has a close friendship with Mary, and though their antics can get in the way sometimes (giggling, running away from class, throwing paperclips, and so on), there was always something fun going on. It was a pleasure having him in my class.

Jonas enjoyed computer class:

Aa Bb Cc Dd Ee Ff Gg Hh Ii Jj Kk Ll Mm Nn Oo Pp
Qq Rr Ss Tt Uu Vv Ww Xx
Yy Zz 1 2 3 4 5 6 7 8 9 10 etc.

It was the first time he made all A's and B's in a school year—usually there were a couple of C's in there. Physical Education was rough because that was the year of his corduroy pants, and he had to run the mile in those pants, as well as doing sit-ups, jumping jacks, push-ups, and the shuttle run. Coach would always say:

"Perfect practice makes perfect."

Mary and Jonas would spend most the time throwing pecans at each other though. She did well in class too, but she always did—all A's.

During recess one day, Mary asked Jonas about his dad. He told her that he was like her, that he didn't have a dad.

"Did he like Advil, too?"
"Don't think so—he died while I was being born."
"How come?"
"The rain."

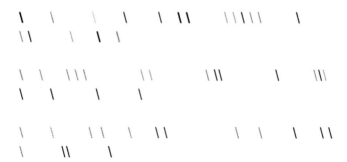

The rain had taken his dad away the night Jonas was born. Mom always said that his dad was the kindest man with the gentlest eyes. She never married anyone after him, and during the thunderstorms, she would stand outside and ask the sky what it would like for dinner.

FIFTH GRADE

Blood

Bryan punched Jonas in the nose. There was blood. And Jonas wasn't crying, but the tears just came out of his eyes—once his vision cleared up, Bryan was still standing in front of him. Jonas didn't hit him back, because he had promised Mom that he would never clench his fists.

"With a closed hand, you can never hold anything true."

Mrs. Rolan came out and told Bryan to go to the

principal's office, where she would meet up with him and Principal Raykes to discuss the situation. As he walked away, Mrs. Rolan put her hand under Jonas' chin and lifted his head.

"Your nose."
"I know."
"What happened?"
"Fuck."
"Go to the principal's office."

It was his first pink slip—a paper of discipline that all students dreaded, because it meant scolding and punishment. Mom didn't punish him though—she was happy he didn't hit back but disappointed that he cursed.

"Where did you learn that word?"
"Something to do with Mary's dad."
"Isn't Mary's dad gone?"
"Fuck."

He ended up getting punished—he couldn't play with Mary during the weekends for a two weeks. For both of those Saturdays, he wrote her letters.

Jonas never knew why Bryan punched him. Everyone else thought it was because he had a crush on Mary, but she didn't like him back, and he was jealous of their friendship. The next day at school, the class stared at Jonas' black and swollen eye. During lunch, Mary put her turkey sandwich against his eye, but it didn't help much, since the sandwich was already warm.

Bryan apologized and gave Jonas his gummy worms. Jonas gave him his cookies to accept his apology. He never punched Jonas again, and one time, he stood up for Jonas when Emily was picking on him. All around, Bryan was a nice guy—the punch made them closer.

Letter 1

Dear Mary,

I got punched! I am punished now and I will see you in school next week. Miss me.

Letter 2

DeaR Mary,

One more week and then we can play on the weekends. Remember to not say fuk in front of your mom.

From Mary:

Hey buddy I wish we could play on SaturDay. We can see other on recess soon. Next time I will not back out and eat the rock.

JONAS, FOR MARY

Chariot: Eulogy For A Paramour

She would come up from behind and put her arm around my neck and press me against her body to give me this hug, this hug, like the world disintegrated and all that was left was us two floating in space between dust and stars.

She would whisper in my ear, as she would place her other

arm around my stomach; I could never understand what she was saying, because she was speaking so gently, like a dying dog, but I would close my eyes and nod, slightly. She knew I couldn't hear her, but she knew I understood her. She was a lullaby. She wasn't a dream. She was just a kite in a vacuum; she was the great disappearing act of the century.

She's gone now. And I'm in bed—in room 802—staring at the wall, making figures against the dull white, out of shadows of my imagination. Trying to get away from it all, but it is all right here, embedded between the intestines of my brain. I see her swimming towards the curtain, in a red bathing suit, with her hair splashing against the Monet painting and the thermostat. She was a horrible swimmer. Doggy-paddled mainly, but now that she's gone, she's like a dolphin.

I can just look at my palm, and see her dancing along the thin crevices, swerving her body around, but never losing control, but always staying on the lines, always cutting on the line, always coloring between the lines.

I close my eyes now, and see ourselves among the dust and stars again. These memories are not as vivid though, not as alive. They slowly diminish into a pile of ashes resting on my stomach. I close my eyes even harder, concentrate, focus, remind myself to buy some Ginko, and clench the blanket so hard, my fingers pierce the cloth.

MIDDLE SCHOOL

Sixth Grade

New school. New surroundings. New people. New clothes. New, new, and new. It was like they were adults, when they walked into the classroom for the first time, trying to impress the teacher, trying to impress their old friends, and trying to impress the ones they didn't know.

Paul clung to Jonas since that first day until the last day of that school year. He went on, on his own, though, when he realized he was who he was, and no one should make him change. Jonas would tell him:

“Be tall.”

Mary. Mary and Jonas hung out during summer before Sixth Grade—he saw her almost everyday. But when she walked into the classroom, it was like he was seeing her for the first time. Mary had all of the boys looking at her. She liked it, too. And every time she smiled, her eyes became ten times shinier. Her breasts were on their way, too, and Jonas, as well as the rest of the boys (except for Craig, Gary, and Harrison) couldn’t help but to look at her chest, which would only get bigger every Spring.

Not as the Tulips Bloom in Spring,
you’d rather wither until Winter.
You are a contrast to the Seasons,
in disharmony, a solace found.

A confusing life you live,
where in cold shadows,
your chest is drawn towards the Sky
like Petals In Sunlight.

A Stem cut from a lonely Garden:

You sing from the freezing depths of your Lungs,
‘Spring is here,’

and you are Gone.

Mrs. Brin would teach her class sitting on a stool, and every time she crossed her legs, from left over right to right over left, there was a collective sigh from the class. It was the first time Jonas saw a lady's underwear: bright white, peeking out from her black skirt, looking at the students as hard as they were looking at it.

Mary was crying in the boy's bathroom:

"How come you're here?"
"I'm crying."
"How come you're crying?"
"The pressure."
"Barometric? I can explain it to you."

She put her hand on his shoulder.

"I'm so glad I know you."
"Do you want me to kiss you?"
"No thanks, but thanks."

Jonas didn't find out from her why she was crying in the boy's bathroom, but the next day, in Science class, she switched her seats, from sitting next to Derek to sitting next to Jonas. Derek kept looking at her, and when Jonas looked at Mary, it looked like she was trying her best not to look back at Derek.

During lunch:

"Did anything happen between you and Mary?"
"What did she say?"
"You two just look awkward."
"She didn't want to go out with me, so I called her a bitch."

Jonas thought about Mom and how she always said to keep an open hand, so he slapped Derek in the face and pushed him, but he only moved back a few steps. He ended up beating Jonas to the ground, and all Jonas could remember was the ringing in his head. It sounded like this:

whirrrrrrrWHirrrrrWHIrrrrrrWHIRRrrrr
WHIRRRRRwhirrrrrWHirrrrWHIrrrrWHIRr
WHIRRRRRRWHIRRRRRRRwhirrrrrr
WHirrrrrWHIrrrrrWHIRRrrrrWHIR
WHIRRRRRwhirrrWHIRRWHIR
WHIRRRwhirrrWHIRRRwhirr

It made a looping sound in his head.

W H I R R w h i r r W h i r r W H i r r W h i r r
W H I R R R R R W H I R R W H I R

"Damn, I love you so much, Mary."

W H I R R w h i r r W H I R R w h i r r r r r r
W H I r r r r r w H i R R R R
w h i r r r r r r r r r W H i R R r r r W H H H I R R R R
w h i r r r r w h i r r r r w h i r r

When Mary finally saw Jonas the next day, she asked him about the fight.

W H I R R W H I R R w h i r r W h I R r W h i r r
W H I R R w H I R W H I R R w h i R r R r r R
W H I R R R w h i R R R R W H I r r r R R R
W H i R R W H I R R w h i R R r r W H i r R R R W h i R R R

"Why Derek?"

W H I R R w h i r r r W H i r R R R W H I R R R
W H I R R R R w h i r r r w h i r r r
W H I R R R W H I R R w h i r r W h i R R R

"He was trying to take my lunch money."

W H I R R R W H I R R R w h i r r r W H I R R R
w h i r r W H I R R w h i r
W H I R w h i r W H I R R R W H I R R R W h i r

Since kindergarten, she knew Jonas' tell. She knew he was lying when he looked directly into her eyes.

"Don't do it again."

W H I R R W H I R R W H I R R
W H I R R W H I R R
W H I R R W H I R R R R
WHIRRRR

Harsh tone.

"You can really get hurt, worse than that."

W H I R R r r r r r r W h I R R R R H R R r r r r r
R R R R r r r r r r r r r r r R R R R R r r r R R R r r r r
R R R R R R R r r R R R r r r r r

"Sorry."

R R R R r r r r r r r r W H i r r W H i r r r W H i r r r
W h i i r r W H I R R r w h i r r r W H I R
W H I R r R r R r R r R r W H I R R R W H i R R r r

"Things like that are going to happen to me a lot, I think. But I'm in control. Don't worry."

W h i r r W H I r r r W h i r R R R W H I R r
W H I r r r m m M - - - - m m m - - - M M M - m m m m
m m m M m - - - M m M m M m

"This is Ground Control to Major Tom."

m m m m m m m - m m m m m - -
m m m m m - - - - - - m m m m - m - m -
m m m m m - - - M M M M M - - m

She made his head ring too. But in a good way. It was more soothing that anything else. It was like someone was humming in his head:

M m m m m - - - - - M m m m - - - - -
M m m m m m m m m - - - - M m m m m - - -
M m - - - - M m m m m m m m - - M m

And so on.

Seventh Grade

The year of the sex education: penises, vaginas, chromosomes, clitorises, tampons, foreplay, missionary, periods, eggs, sperm, erections, secretions, giggling, pointing, shouting, tapping, staring, vomiting.

Jonas didn't pay attention at all. He failed the quizzes and tests. He was too busy thinking about how to make the tennis team. His serve was never good, and his movement was more stagnant than mobile. Didn't make the team in Sixth grade, but Jonas worked hard all summer, and he was getting ready to try out for the tennis team. While the rest of the class was learning about the how the pieces fit between two bodies (insert penis here), he was thinking about Andre Agassi.

It was safe to say that everyone thought that Mary was the prettiest girl in school. However, she had yet to say yes to going out on a date. Jonas' friends told him that he should ask her out.

"Why? We go out all the time."

"But in a different way. In way that you can stick your tongue into her mouth."

He had thought about it before, and that was the extent of his own sex education. But Jonas never acted on it. Somewhere inside of her, he knew she knew what he thought about her. She could tell by the way he didn't look at her in the eyes. But it was more like a silent agreement or understanding, and they left it at that for a while.

Didn't go to any of the school dances—Jonas was working on his serve, practicing hitting the ball against the side of Mom's house, while the class was at the gym, learning where to put their hands, how to move side to side, trying to get some kind of vibe, hoping that at some point, lips will be able to touch, bodies will be able to be touched.

Jonas didn't make the tennis team in the end, and that was his last attempt at sports. Mary was on the team—one of the top players on the team. He would watch her play, cheer her on. In between sets, she would look his way and smile or wave.

In sync with sex education, Jonas had two strands of hair growing from his testicles. He showed them to Mary.

"They're pubic."
"I have some, too."

She had much more than Jonas.

"Mine are longer though."
"You should name them."
"Laurel and Hardy."
"I have too many to name, but the first one I had was named the Violinist."

He showed her what he had underneath his arms.

"Not bad."

She didn't have any under her arms.

"I win."

She put Jonas in a headlock, and then he picked her up and they both fell on the floor of her bathroom, naked and laughing.

Quiz 2

After watching the video on sex education, please answer the following questions. Complete sentences must be used in your explanation.

1. Can only males masturbate? Explain your answer.

<u>Well, I think the problem was that I was throwing the ball too high and too much in front of me. I need to keep my elbow in, and keep my eyes open, making sure the racket properly hits the right side of the ball.</u>

2. What is premenstrual syndrome?

My lateral movement needs work—it's difficult, because my feet get tangled up. I think that if I practice my plyometrics, that I can help to better this situation. Also, I need to be aware of all of the different areas of the court.

3. What is the egg? What is sperm? Be descriptive in your response.

Andre Agassi had a powerful two-handed backhand—I'm striving to have the same technique. Anticipation is also key when returning a hit. The opposing player must be studied. Not only keeping an eye on his eyes, but also his body motion and positioning.

Eighth Grade

The class went on a ten day hiking and camping trip. It became more of a love fest—starting from the bus stop on the way to the wilderness and ending on the bus strip on the way back home.

Laney hooked up with Ronald; Carrie hooked up with Thomas; Laney also hooked up with Steve; Suzy—Bryan; Kerry—Angela, and the list went on.

Jonas didn't hook up with anyone. He didn't know how to. He didn't know he should have been trying. He was too busy trying to keep his socks dry and his toilet paper abundant. Mary was fingered for the first time—Daniel. Jonas was curious:

"Was it neat?"
"I think so—it felt good."
"How long does it last?"
"Felt like forever."
"Is it something I can do?"
"As long as you have fingers."
"Maybe one day."
"You'll make someone real happy one day."

He didn't tell Mary how jealous he was of Daniel, more so because it was her first time, and she didn't talk about it with anyone else, except him. She needed someone, he was the person. He would always be the person. Daniel was a nice guy—shaggy blond hair, brown eyes, and one of the best baseball players on the team. Jonas was happy for both of them.

The first day of school, back after the camping trip, the campus was full of gossip: who kissed who, who didn't kiss who, how, when, where, and why.

JONAS, FOR MARY

Carving The Air

You know how when we were younger, we would look at the clouds and see different things like dinosaurs or turtles or cars and all? I could never see any of those things. I never saw it. I tried hard too. I'd concentrate and concentrate but could never see what everyone else would see.

Not too long ago, in a galaxy far far away, I was sitting on a roof of someone's house, someone I didn't know, but the roof looked like a nice place to sit. I climbed up an oak to see if I liked it, and I liked it, liked it so much that I sat there every Sunday like I was going to church. I was sitting there staring at the clouds trying hard to see something concrete, but as the clouds separated, I saw the moon. It wasn't anything special or anything, it looked quite dull and somewhat broken, but as I chained smoked I saw a blinking star directly above the moon. As I stared and stared at it, I didn't see a dinosaur or a car or anything, but I saw someone dancing on this shiny little thing in the sky. I saw a white dress wavering, flowing, swerving, and long black hair, so black, it made the night look bright, and her lips, lips that made me want to kiss her chin because no one should ever receive such a gift as to touch those lips, and I couldn't see the rest of her face or anything, but damn, was she a great dancer. She jumped from star to star, planet to planet—an acrobat of the universe, and she moved her body up and down, left and right, like she didn't need oxygen, but oxygen needed her. I think she even turned her head in my direction and waved, but that could have just been my imagination. She gradually drifted away as she danced from one sparkling diamond to another, and an infinity later, all I saw was the cracked moon again.

I've gone back every Sunday since, looking for this astrological performer, but I haven't seen her yet. Sometimes I dance on the roof myself, trying to remember her body from such a distance, and I'm lucky the owners of the house have yet to notice cement feet, but I've become a better dancer, flowing a bit better, more in harmony with the air as I try to carve it with my body. And I have a new plan now. Next Sunday, I will really try to look for her, not like look for her while sitting on the roof and staring hard into space, but I saved up enough cash and bought some athletic shoes, and I'm going to try to jump from the roof to the moon, and then from the moon to star to star to star to planet to planet, until I find this lunar dancing spectacle, so I can ask this lady for a dance. Who knows—maybe I'll meet her on Saturn or Jupiter or maybe she'll be right next door, on Mars. Just one dance in the sky, and I'll come back down and never dream again.

HIGH SCHOOOL

Ninth Grade

He didn't like memories—whether they were good or bad. Thinking back. Looking back. These events, these snapshots embedded in their brains, tucked in, underneath our skin, they can be burdensome. Why remember? Why remember the feeling he can never have again? It was a torture of some sort, a personal torture that can only be cured by amnesia. Nostalgia, itself, is a disease that spreads from the mind down to the stomach, and sometimes to the legs, causing them to shake. With all that being said, Jonas wouldn't have it any other way.

They were newly born adults again. They crossed their

legs and drank coffee and wore button up shirts. Mary sat in the next row, a couple of seats ahead in History class. She was sophisticated. Everyone still stared at her breasts, including Jonas, and she finally gave in, and started dating someone named Michael. He was okay—nothing special, but Jonas knew he wouldn't do any harm to her. Mary and Jonas—they didn't hang out as much. Getting into that phase in their lives, where other interests and other people kept them occupied. Well, at least for her. Jonas didn't do much differently than before. He did, however, have his first crush after Mary. It wasn't like he didn't like Mary that way anymore, but more like, he knew nothing would happen, at least then, so he kind of kept his mind open. Her name was Jennifer.

She was a quiet lady. Nice to everyone, including those that no one else really paid attention to. She sat behind Jonas in History class and would kick the bottom of his chair on accident from time to time, which was a nice excuse for Jonas to turn around and look at her.

"Quit kicking me."

"So sorry—I was just trying to shift my body."

"Quit shifting."

"I'll try to stop."

"I like your scarf."

"Thanks."

"Is it Parisian?"

"It was made in China."

"They make great scarves."

Mary would turn her head to look at them whenever Jennifer and Jonas would whisper to each other. His intentions were never to make Mary jealous, she knew how he felt about her, but it eventually got in the way. His interest in Jennifer didn't suit Mary, much:

"Why Jennifer?"
"She's nice."
"I'm nice. There are a lot of nice girls out there. Why her?"
"I like her voice. And her skirts."
"Do you like my voice?"
"It's the best I've ever heard."
"Just be careful."

Voice
Voice
Voice

Voice
voice
Voice
Voice
Voice
Voice

It was an echo that would never go silent in his mind—Mary's voice. There were an infinite amount of voices out there, but it was Mary's that always made its way to his ears and stayed there forever.

"Just be careful."

Jonas didn't realize until later that she was looking out for him. She wanted to make sure that Jennifer wasn't going to mess with him. Mary knew the power of a woman's mind and body. She used her own sparingly, but she knew that others may not be able to control it as well as she did.

dance dance dance dance dance

dance

dance dance dance dance

dance dance dance

dance dance dance dance dance

dance dance dance dance

Jonas' first attempt of going to a school dance consisted of asking Jennifer if she would accompany him. It was for Winter Formal.

"I was wondering.
I was wondering.
I was wondering if you would like to go to the upcoming dance with me. I would love to take you."
"That's sweet, but I'm waiting on Jeff to ask me."
"I understand."

And that was that—his first and last attempt of asking someone to go to a dance with him. He didn't go to Winter Formal. He didn't go to any dances until Senior Year, for Prom.

Mary found out that Jennifer didn't want to go with him, or that she was waiting on someone else to ask her.

"You okay?

At least you tried.

Told you she was no good. Who wouldn't want to go

with you?"

Jeff did end up asking her to go to the dance, and they hooked up at the after-party. Since then, whenever Jennifer would kick the bottom of Jonas' chair, he didn't turn around anymore. She kept kicking and kicking, but he remained still.

"Why don't you turn around anymore when I kick your chair?"

"I'm not Jeff."

We, The Sophomores

The softball team won the state championship. Jonas didn't go to any of the games, but he heard that they won by at least five runs in every game they played. It was the best team the school had in history, in any sport.

Mary had a new boyfriend—he was from another school. Jonas didn't like him. He didn't like the way he slicked back his hair or the way he grabbed his crotch to adjust, or the way he talked to Mary, or the way chewed his gum. He didn't like him.

They were all hanging out at The Teapot, a 24-hour

restaurant, which served waffles and hamburgers. Jonas threw up in the bathroom there, once, after drinking six successive cups of coffee, a Sunkist, and chain smoking stale cigarettes he stole from one of his teachers. The new boyfriend, Robert, was sitting next to Mary, and Jonas was sitting across the table from him. He kept on talking about the cheerleaders from his own school, and Jonas could tell Mary was getting frustrated by the way she kept looking around the restaurant, shifting away from him, and attempting to change the subject.

"Marla has a fine ass. How come you're not a cheerleader? You have a nice ass, too."
"Too busy with school and being on the student committee."
"Student committee. Stupid."
"It's not dumb."
"I'm surprised you're so popular not being a cheerleader and all. Usually I date cheerleaders."

Jonas spat on Robert, and Robert beat Jonas up in the parking lot. He left, leaving Mary with Jonas.

"Why did you spit on him?"
"How the hell do you like that guy? Tell me."

"I like the way he looks at me. And he's funny. He's just different when he's around people."

"By different, you mean obnoxious and socially inept."

"He's nice."

"I don't like him."

"Never said you had to."

"Why are you going after guys after that? You don't need to, you know. You're already popular."

This was their first and only argument.

"Is that what you think? I'm trying to be popular."

"It doesn't make sense, then."

"Why are you being so mean?"

Jonas felt horrible.

"You deserve to be beaten up."

"I did it for you."

"Stop doing things for me. I don't need you to do things for me."

She didn't let Jonas take her home that night. She called another friend to pick her up. Jonas didn't leave until the friend came, though.

Of course, they apologized to each other later on. She

had the same feeling of emptiness in her stomach that he did. She couldn't sleep that night just as he couldn't.

"Sorry."
"Sorry."

She broke up with Robert a few weeks later.

He cried.

She didn't.

"Can I tell everyone that I broke up with you instead?"
"Say whatever."

RAH RAH RAH!
RAH RAH RAH!
RAH RAH RAH!
RAH RAH RAH!
RAH RAH RAH!
RAH RAH RAH!
RAH RAH RAH!

Mary ended up trying out for the cheerleading team, and once she found out that she made the team, she opted out of being a cheerleader.

"I have a nice ass."

"Yes you do, Mary. You have a nice ass."

Regina offered fellatio to Jonas—it was the first time anyone asked him to partake in this sexual maneuver. He got nervous and said no. They were at a party. It was also his first party. Gerald's parents were out of town for the weekend, and he went with Mary, just to keep her company after the break-up with Robert. Regina walked in while Jonas was urinating in the bathroom—all of this could have been avoided if the lock worked, or if she would have knocked.

"I'll have a go."
"I'm sorry?"
"Since it's already out, I'll have a go at it."
"The bathroom?"
"Your penis."
"I'm fine, thanks."
"Your loss."

Perhaps, it was his loss. Perhaps, it would have opened him up to a whole new world—making him more social, making him feel more like a man. He wasn't ready yet. He was still working on trying to hold hands and kissing someone on the cheek.

Mary was amused.

"Regina offered me fellatio."
"A blowjob?"
"That too."
"And?"
"I politely declined."
"Good."
"Why?"
"I don't know."
"You don't like Regina? I think she's nice."
"Of course you think she's nice—she's willing to put your penis in her mouth."
"I should have said yes."
"When it happens, it will happen."
"Yes, Zen master."

He asked Mary if she ever performed fellatio. She said yes. He asked her if she liked it. She said that she got used to it, and that it got her set on oral fixations. It was why she started smoking.

"If you want, I can give you one."
"Really?"
"Better me than anyone else."

"True. You're not going to bite down on it are you?"
"Depends if you're nice."
"Maybe later."
"Just let me know."

He ended up on top of the roof of the house with Billy.

He was playing the guitar and smoking pot.

"Want a drag?"
"Maybe later."
"Do you play guitar?"
"Just the A chord."
"That's a good chord."

<u>Playing The A Chord</u>
1. Put your index finger here.
2. Then put your middle finger there.
3. Then put your ring finger there.
4. Press down hard. (It may be hurt in the beginning.)
5. Strum your soul away.

Billy threw his guitar and it broke as it landed in the front yard. Then he jumped off the roof and broke his ankle. Jonas climbed back through the window.

The Juniors

Tractor had a strong headlock on Jonas. He called Mary a bitch for not putting out, which meant that she didn't let Tractor feel her up. Jonas mouthed off to him, not really knowing what he said, but it was something that Tractor didn't like and he took it personally. There was no way out of it either. Perhaps, he could have tried hitting him in the groin, head, or stomach, but he used words instead.

"Why are you headlocking me?"

"You called me a piece of dumbfuck, and you said you

fucked my mom."

"Come on, you know I really didn't sleep with your mother. But you really are a piece of idiotic fuck."

His grip tightened. He heard Mary in the background, cursing at him for starting something with Tractor.

Tractor.

No matter how small you type his name or how low you say his name, there is still so much power behind it.

Tractor.

That's more like it.

The P.E. Coach, Mr. Williams, broke up the fight. He didn't give them detention, but he told them that if he caught them fighting again, they would have to run laps around the gym. Tractor made the wrestling team, too.

Jonas went more and more into solitude. He offered Maggie three dollars if she moved away from a bench, so he could eat his burrito alone. She agreed.

Mary and Jonas rarely talked. She grew more and more popular. She knew it too, and made a special effort to include Jonas in her outings or groups, but they both knew that it wasn't a good idea. However, he did appreciate her effort.

They did go see a movie one night. For the first time, Jonas was nervous being around Mary. She was the same person, at least, when she was around him. But he wasn't the same around her anymore. The crushes he had for her had slowly turned into love.

"I love you."
"I love you, too."
"Will you marry me?"
"Of course I will."

And he gave her a dandelion. He wondered if she ever knew that he was really asking her to marry him.

In the movie theater, Mary crossed her legs, facing Jonas. It was his first erection.

"Oh really, Mister?"
"Sorry."

He went to the movie theater bathroom and masturbated for the first time—didn't take longer than eight seconds.

Who knew my own hands could cause such happiness?

"You're all good now?"
"Right on."
"Happy to help."

She punched him in the shoulder.

They finished the movie and she took him home. They just talked. That was what was so neat about Mary. She didn't make Jonas feel embarrassed. She made everything seem natural.

The next day at school, they didn't get a chance to talk to each other, but they did catch each other's eyes during break.

They were looking at each other.

Gazing.

He was lost, again, like he was in Second Grade. Mary mouthed something to him, but he couldn't tell what she was saying. He just kept looking into her eyes from twenty to thirty feet away.

When she's happy, her eyes would look watery, like she

was crying. It would make him want to cry. It was like her eyes were her mouth, and they would speak to him, like that Let It Be song, *"whispering words of wisdom."*

If he had a smile for every time he looked into her eyes.

Sometimes he wondered so hard about what she was thinking, about what goes on in her mind.

"Take my hand."

She took it.

"Now what?"
"I'm trying to read your thoughts."
"Like a palm reader?"
"Through your eyes."

They stared at each other.

"Then why did you ask me to take your hand?"
"Because of your soft skin."

She slapped him on the back of his head.

"So could you read my thoughts?"
"Yes."
"What was I thinking?"

"You should know shouldn't you—they are your thoughts."
"I want to see if you're right."
"I am."
"How do you know?"
"Because you took my hand again."

JONAS, FOR MARY

Brief Encounter In The Museum

I hear the calling of my name in the sound of footsteps; I do not stop until I feel a tap on my shoulder. I turn around and see pain's glory; we do not speak, nor do we continue to walk, but we stand still and study each other until we fall.

We are motionless: I am a mosquito, you are amber, and we're fossilized.

Millions of years later, paleontologists will find us
on a piece of bark; they will see the petrified look
on our faces and wonder what had caused such eyes.
We cannot answer, for we are in each other's mouth,
wishing we were made of cotton and orange peels.

He wished that would have happened.

One evening, Mary attempted to teach Jonas cunnilingus.

"You need to know these things, she said to him. Trust me, it will go a long way with whoever you're with."

SEX EDUCATION

A One-Act Play, A Tragedy

Characters

Jonas:
A young adult male, who, at the age of 18, is trying to figure out how it all works.

Mary:
A young adult female, who, at the age of 18, is trying to help Jonas to figure out how it all works.

Time
Present—4 P.M.

Setting
Mary's bedroom.

Mary and Jonas are in bed—Mary sits with her back against the headboard. Jonas is at the other end of the bed, on his stomach, between Mary's legs.

JONAS
Where is it?

MARY
It's right there.

(*She points between her legs*).

JONAS
Like, the whole thing?

MARY
No, just that part right there.

(*Points*).

JONAS
I'm not understanding.

MARY
There's not that much to understand.

JONAS
I don't see it.

(*He strains his eyes*).

MARY
I'm sorry.

JONAS
Don't be sorry—I'm sorry. I wish I could see it.

(*He scratches his head and rubs his eyes*).

MARY
Here, you see.

JONAS
Maybe. Now I'm like seeing everything in a blur because I've been looking so hard.

MARY
Blink.

JONAS

(*Blinks*).

No good.

MARY
Well I don't know what to do.

JONAS
I really don't know what to do. Can you put some more light on it or
something?

MARY
Every light is on in the house, sweetie.

(*Mary adjusts the lamp*).

JONAS
Snap.

MARY
Yeah, snap, look harder, it's that bit right there.

JONAS
I don't think it's going to work.

MARY
Why don't you just try it out?

JONAS
No way. I want to know what I'm doing.

MARY
That's sweet.

JONAS
I'm a bit scared.

MARY
Don't be.

JONAS
Here I go.

(*He closes his eyes*).

MARY
That's my thigh, dear. You can keep your eyes open.

JONAS
Snap.

MARY
Yeah, snap.

JONAS
Can I take a break?

(*He stretches his back*).

MARY
Sure. I'll get it pierced or something, and maybe that'll help.

JONAS
Sorry I'm so lost.

MARY
Don't be.

JONAS
Do you still love me?

MARY
Of course.

(*Mary runs her hands through his hair*).

JONAS
Thanks.

MARY
Well I needed to go run some errands anyway.

JONAS
Yeah?

MARY
Yeah.

JONAS
Well call me later tonight. I'm going to go eat some carrots.

(*He stands up and walks towards the door*).

MARY
Okay, Dear.

END

They saw themselves as the royal ones—the ones who had been through it all. They could walk around and not notice each other, because there was nothing else to prove. There were still groups—the cool ones, the athletic ones, the smart ones, the quiet ones, and some of them blended with each other to form a sturdy Venn Diagram.

Mary was single during all of Twelfth Grade. Jonas

didn't know why—they never talked about it.

Although he couldn't successfully please Mary, he did unexpectedly lose his virginity to her during the Senior Trip. They were in France, and he was in her hotel room—they were talking about how it would be neat to fluently learn another language.

JONAS, FOR MARY

Parlez-Vous?

"I don't know any French," I said. "You only need to know one word mon cher," she said, "Amour." "Oh, okay," I said, "What's that? Is that your name?" She moved her hands, removing my body from reality, making an atheist feel holy. She closed her eyes, and I couldn't keep mine open, and in the darkness, I saw her touches, each touch, each graze, giving a little glow, like a million lightning bugs floating around us. I saw her moans, and I kept muttering to myself, "Merriam-

Webster, Merriam-Webster, Merriam-Webster," hoping that the next morning I won't forget to buy a French-English dictionary.

After they were done, they just lay in bed staring at the ceiling. She fell asleep with her head on his chest, and he made sure not to make any kind of movement because he didn't want her to wake up. He wanted her to be on his chest for as long as possible. He didn't know what he was doing—she did. She guided him, and after all the movements were finished, she kissed him on the lips. He didn't know what to say. He didn't say anything. He just panted heavily, trying to remember every second of what had just happened.

Sr. Prom. She wore a long dark green dress, tight against her body. Her hair was fixed in a way Jonas had never seen before, making her look more sophisticated and a bit unapproachable. Mary made Jonas go. She said it was important that he went, and that it was important that he went with her. At the dance, she moved her body to the music like a stripper. All eyes were on her, and the chaperones didn't know whether to tell her to stop dancing like that or to continue watching her, taking notes, so that they could use these moves on their loved ones. She grabbed Jonas and pulled him to the center of the room, giving him some kind of lap-dance, though they were both standing. One of the teachers finally came and told them to cool down. He went to the restroom for a few minutes and when he came back out, he saw Mary sitting down.

"Resting your legs?"
"Are you having a good time?"
"Very much so—thanks."
"You'll never forget me, right?"
"Why do you ask?"
"I don't know—there me be a time when I won't be around anymore, and I just want to make sure."
"Well, the only way I can forget you is if someone took my head away."

He didn't go to graduation. She did. She kept calling him the night before, but he didn't answer the phone. He just sat on his bed and cried. The end was approaching, and he wasn't ready. Mary was hinting at something at Prom, but he didn't know what she was exactly talking about. He had a feeling that he would soon know.

The day after graduation, he received a letter from Mary. It was a goodbye note—saying everything he wanted to hear—she loved him, they were the closest of friends, she wished she could have stayed around longer to live the rest of her life with him. But.

But.

But.

But.

She used the same kind of pills her dad had used. Twelve of them. She was found underneath her bed, like she was hiding from something.

o

ooooo

oooo

oo

JONAS, FOR MARY

i am here And You Are Gone

My head rests with drooping brow, with closed eyes, with breaths full of knots and splinters. Oh where, oh how they fade away, these stars I try to catch—catch a star, catch a star. In this pocket of mine I have a jar full of jelly beans—I took them from a dream I will have tomorrow; I will not eat them; I will not smell them or throw them like rocks. I will keep them in my pocket for as long as I can, because jelly beans are my dreams, my spectrums in a world full of cement blocks. I can never step on a grave. I'm afraid. I'm afraid I will fall in and know the bones beside me. Will it be me, myself? Resting next to my own body—mirrors can be so confusing sometimes, and it's a grave situation. Sometimes I don't know that I am alive. Such gravity and such gravity pulling down on my back, stretching my skin, peeling, revealing spinal cords and strands of sinews inside of me. When once I was a child, I thought to myself, I will be where I need to be. I realize now, where I need to be, is nowhere to be seen. I smell stones, I feel the pins pricking my pores, but when I stick out my arms and hands, and reach for whatever is there, I feel nothing. Can there be? Can there be two eyes looking at me? I will never ask myself this question, because I know the answer, and if I didn't, I wouldn't want to know. I heard breathing once, hot, soothing breathing—its warmth caressed my nose, tickled my nostrils, reminding me of once when I was made of two. Forgetting is not so easy, when you're trying so hard not to remember. Remember? Remember when we would look at each other and say nothing at all. Because saying nothing at all was all we had to say. The words, the vocals, the semantics, tones and pitches were all found in a hidden place in the space between us. And when I turned around, and then turned around again,

that space, that little nook of our thoughts was taken away.

Just like that—either you were gone, or I haven't left yet. I will never turn around again; I don't want to miss what will be behind me anymore.

I don't know if you're just bones in the ground, like Mrs. Carrados, or if you're somewhere in the air, bits and pieces of particles, a halo floating around my head. Fragments, like these others, here, in this bed, these gathered remnants of who we were and what will never be.

And I will know soon where you went, and what will happen next.

Mary Mary Mary Mary Mary Mary
Mary Mary Mary Mary Mary Mary
Mary Mary
Mary Mary Mary Mary
Mary Mary Mary Mary
Mary Mary oooooo oooooo Mary Mary
Mary Mary Mary Mary
Mary Mary Mary Mary
Mary Mary
Mary Mary Mary Mary Mary Mary
Mary Mary Mary Mary Mary Mary
Mary
Mary
Mary
Mary
Mary
Mary
Mary
Mary
Mary
Mary
Mary
Mary
Mary

Breinigsville, PA USA
10 November 2010
249106BV00001B/133/P

9 780976 346074